AF606946

AVIATION

B-26 Marauder

Martin's Medium Bomber in World War II

DAVID DOYLE

Library of Congress Control Number: 2018957914

Designed by Justin Watkinson
Type set in Impact/Minion Pro/Univers LT Std

ISBN: 978-0-7643-5664-3
Printed in China

Published by Schiffer Publishing, Ltd.
4880 Lower Valley Road
Atglen, PA 19310
Phone: (610) 593-1777; Fax: (610) 593-2002
E-mail: Info@schifferbooks.com
www.schifferbooks.com

Acknowledgments

As with all my projects, this book would not have been possible without the generous help of many friends. Instrumental to the completion of this book were Tom Kailbourn, Rich Kolasa, Dana Bell, Stan Piet, the late Roger Freeman, Scott Taylor, Brett Stolle, and the staff of the National Archives. Most importantly, I am blessed to have the help and support of my wonderful wife, Denise, for which I am eternally grateful.

Contents

	Introduction	004
CHAPTER 1	B-26 and B-26A	005
CHAPTER 2	B-26B and B-26C	032
CHAPTER 3	B-26F and B-26G	074
CHAPTER 4	XB-26H	096
CHAPTER 5	Production	100

Introduction

In March 1939, the Army Air Corps issued Circular Proposal 39-640, setting out the criteria for a desired new medium bomber. Seven firms responded, including notably North American Aviation, which offered their model NA-40 (the aircraft that ultimately would become the B-25), and the Glenn L. Martin Company, which offered their Model 179—which became the subject of this book, the B-26.

Unlike the typical Air Corps procurement procedure of the era, which involved the evaluation of an actual flying prototype, the Air Corps responded to the proposals received by issuing a contract to Martin to build 201 B-26s (and to North American for 184 B-25s). Thus, there were no experimental and evaluation aircraft of the XB-26 or YB-26 series.

The B-26 was powered by a pair of Pratt & Whitney R-2800 Double Wasp radial engines, which were rated at 1,850 horsepower each. These would move the B-26 at 315 mph at 15,000 feet and provided for a bomb load of 4,800 pounds.

On November 25, 1940, a B-26 lifted into the air for the first time. While the first four production aircraft were used for flight testing, the first operational unit equipped with the new bombers was the 22nd Bomb Group, which began to be equipped with B-26 aircraft in February 1941. The high wing loading of the early B-26 in particular meant that while the aircraft were fast, skillful handling was required, especially when landing. Thorough training was a must.

The 22nd Bomb Group would also be the first to use the B-26 in combat, striking Rabaul on April 5, 1942.

CHAPTER 1

B-26 and B-26A

Unlike many US military aircraft, where the base model number was merely a generic term for the type, in the case of the B-26, the initial production model was in fact designated "B-26." On August 10, 1939, the initial order for 201 B-26 aircraft was placed. Just over a year later, on November 25, 1940, William K. Ebel, serving not only in his position of Martin chief engineer but also test pilot, took the type for its first flight. That flight, in aircraft serial number 40-1361, was uneventful, but problems with the nose landing gear arose on subsequent early-production aircraft. Problems led to a rapid redesign of the front strut, before the true—and very simple—cause of the problem was discovered.

Martin, it seemed, was to deliver the aircraft without armament. The lack of weaponry naturally affected the aircraft's weight and balance. Not surprisingly, Martin was award of this, and so they packed the spare parts and tools that were to be shipped with each aircraft in such a way as to compensate for the lack of armament. Air Corps personnel on the receiving end, who were unaware of this careful effort, unloaded the aircraft but did not install the armament. This led to excessive loads being placed on the nose gear during subsequent flights, at least until the problem was identified.

Straight B-26 production came to an end during October 1941, following the completion of the 201st B-26: serial number 40-1561.

That aircraft was followed immediately by the first of 139 B-26A aircraft, serial number 41-7345. Not surprisingly, given that the aircraft went straight from drawing board into service use, some improvements were desired. These aircraft differed from the previous model most obviously by having all the .30-caliber machine guns of the B-26 replaced with the much more formidable .50-caliber machine gun. Less obvious, but at least equally important, was an increase in armor protection. Internally, the aircraft was equipped to carry an auxiliary fuel tank in the aft bomb bay. These aircraft were also equipped to drop torpedoes.

The Royal Air Force received seventy-one B-26 and B-26A aircraft, becoming the first foreign nation to use the type. The aircraft, designated Marauder I, were assigned RAF serial numbers FK109 through FK160 and FK362 through FK380. Seven of these were lost in transit, and five before delivery.

The Martin B-26 went directly from design and mockup stages to production, without the benefit of prototypes, which was an unusual sequence of events for that time. The first B-26 to be completed, B-26-MA, US Army Air Corps serial number 40-1361 and Martin aircraft number 1226, as seen here, took to the air for the first time on November 25, 1940, with William K. Ebel, Martin's chief engineer, in the pilot's seat. This aircraft was used, along with several other early B-26s, for flight testing. *Air Force Historical Research Agency*

The first Martin B-26-MA runs up its engines in a photo dated November 28, 1940. This plane's first flights were out of the Glenn L. Martin Airport, outside Baltimore, Maryland. No top turret or armaments were on this Marauder during the test phase. In this photo, there was a device, probably test equipment, mounted above the right pitot tube. *National Museum of the United States Air Force*

The first B-26-MA is poised on the runway at Glenn L. Martin Airport on November 28, 1940, three days after its initial flight. Martin conducted testing of the B-26 for several months before the Army Air Corps commenced its own flight testing of the bomber on January 28, 1941. *National Museum of the United States Air Force*

The early-production B-26s had spinners on the propellers, but these were deleted by the time the B-26B-MA was in production because the spinners restricted the airflow that the engines required for cooling purposes. *National Museum of the United States Air Force*

Although the Army Air Corps' tests of the B-26-MA began in January 1941, the first B-26-MA had Air Corps markings, including national insignia on the wings, "U.S. ARMY" on the bottoms of the wings, and red, white, and blue stripes on the rudder. The B-26-MA had two adjoining bomb bays, forward and aft, each with its own set of doors. Each aft door was fitted with a window, as seen here. *National Museum of the United States Air Force*

The third B-26-MA produced, US Army Air Corps serial number 40-1363, is parked at Patterson Field, Ohio, in March 1941. Like the first B-26, this one lacked the top turret, the opening for which was covered with aluminum, faintly visible above the star insignia on the fuselage. The code 44/22B was marked on the dorsal fin, signifying aircraft 44 of the 22nd Bombardment Group, to which this plane was assigned in February 1941. Flexible machine guns were mounted in the nose and in the tail. *National Museum of the United States Air Force*

B-26-MA, serial number 40-1363, is viewed from the front left quarter. The propellers were Curtiss Electric propellers, a type that proved problematic in the testing and early operational service of the B-26s. The clear nose was shaped from two sections of Plexiglas with a horizontal joint between them. A machine gun was in a ball mount on the top half of the nose, and a flat aiming pane for the bombsight was on the lower half. *National Museum of the United States Air Force*

In a photograph predating March 31, 1941, four completed B-26-MAs are on a hardstand at Martin's Middle River, Maryland, plant. The plane in the right foreground has received its top-turret dome and a faired-in, windowless tail stinger of the design that would start appearing with the B-26B-MA, while the two Marauders in the left background have tail stinger positions with clear panels but lack the top turrets. *National Museum of the United States Air Force*

The early B-26-MA with the shape of a tail stinger faired in is viewed from the left side. This shape represented the early turret, minus the clear canopy, which mounted a hand-operated .30-caliber machine gun, located above the tail cone. The Plexiglas dome of the top turret appears at this point to have been devoid of internal structures. Note the streamlined fairing over the engine exhaust on the side of the engine nacelle. *National Museum of the United States Air Force*

Two air crewmen, apparently an Army Air Corps pilot and a Martin employee, go over a checklist while uniformed Army Air Corps officers stand in front of engine 1 of an early B-26-MA. The plane is in highly polished natural-metal finish. The earliest B-26-MAs were delivered to the Army in this finish; later, B-26s would be delivered painted in a camouflage scheme of Olive Drab on upper surfaces and Neutral Gray on lower surfaces. *Air Force Historical Research Agency*

Among the several early B-26s parked at the Martin factory, the bare-aluminum bomber in the background has markings for aircraft 15 of the 22nd Bombardment Group. To the right of that plane is a Marauder painted in Olive Drab over Neutral Gray camouflage. *National Museum of the United States Air Force*

Camouflage-painted early-production B-26s are lined up at Martin's Middle River plant. Affixed to the vertical tails are rudder locks, to prevent wind damage to the rudders and their mountings. The rears of several of the Marauders are visible, and they have a type of tail machine gun position that was experimented with early on in Marauder production, featuring a tapering clear canopy that extended down to the tip of the tail. Dorsal turrets are not installed. *National Museum of the United States Air Force*

A photo of two civilian workers next to an early Martin B-26 includes details of the open doors of the escape hatches above the cockpit, the design of the clear nose and the nose landing gear, and the wavy, "soft" demarcation between the upper (Olive Drab) and the lower (Neutral Gray) camouflage paint. Several early B-26s crashed, with the loss of lives, after the pilot's escape hatch came loose and hit the vertical stabilizer. *National Museum of the United States Air Force*

A B-26 runs up its engines at an unidentified airfield. A small, white letter *X* is marked on the fuselage above the nose gear door. Note the horizontal seam between the top and the bottom halves of the clear nose. Parked to the left is a Lockheed P-38 Lightning. When delivered to the Army, the earliest B-26s lacked weapons, radio equipment, and other materials installed by the Army, so the planes were nose-heavy, leading to some nose gear failures until the struts were strengthened and properly distributed loads were installed. *National Archives*

An early B-26 in flight lacks its machine guns. The top turret is turned toward the photographer, with the slots for the guns being visible. A pitot tube is on each wing, which was a feature of the B-26 series until they were deleted starting with the B-26B-55-MA production block in favor of a single pitot tube in front of the nose landing-gear bay. *National Museum of the United States Air Force*

After the first B-26-MA, army serial number 40-1361, completed its factory flight-testing program, it was sent to Laughlin Field, Del Rio, Texas, and used as a training aircraft, receiving the nickname "Gran' Pappy" and a coating of camouflage paint. By the time this photo was taken on November 28, 1944, the plane was near the end of its career, with paint peeling off and the left engine missing. Note the clear dome on top of the fuselage and the "NOT TO BE FLOWN" warning between the cockpit and the wing. "Class 26," marked to the front of "Gran' Pappy," means that this plane has been assigned salvage status as a mockup. *National Museum of the United States Air Force*

The first B-26-MA, "Gran' Pappy," ended its career as a mockup aircraft for training radio-mechanics students at Truax Field, near Madison, Wisconsin, in the latter part of World War II. Here, Pvt. Robert J. Grillhoesl, a radio student, poses in the left cockpit hatch of "Gran' Pappy." The nomenclature stencil indicates that the plane had been redesignated an RB-26, the R prefix standing for restricted. *National Museum of the United States Air Force*

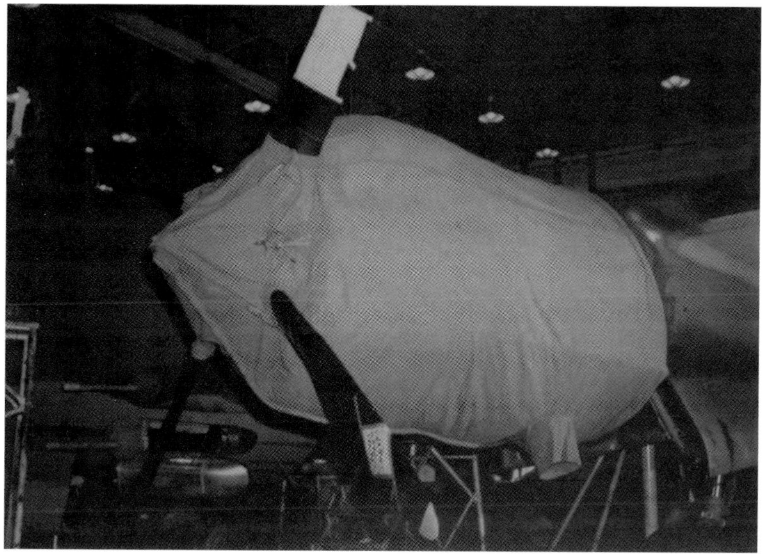

Fitted over the propeller hub and cowl of a B-26 is a fabric cover, most likely part of a winterization kit to prepare the plane for service in a harsh northern climate. Package guns are visible on the side of the fuselage, to the lower left. *National Museum of the United States Air Force*

A mechanic runs what appears to be an electrical cable from the front of the left main landing-gear bay of a B-26 that is apparently undergoing tests with a winterization cover over the left cowl. A sign taped to the propeller blade is obscure but evidently contains a warning concerning not turning the propeller when the plane is prepared for storage. *National Museum of the United States Air Force*

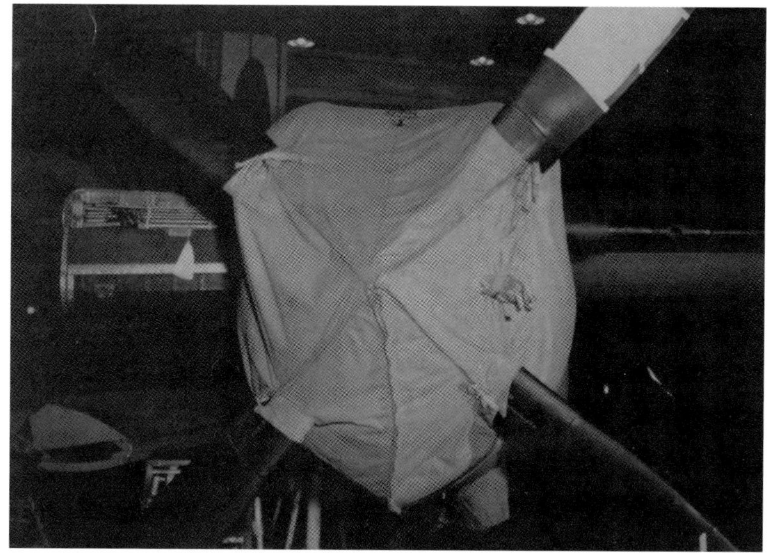

The cover on the left cowl of a B-26 is observed from the front. At the top of the cover is an illegible word over the word "CENTER," with an arrow under it. Deicing boots, for preventing ice buildup on the leading edges of the wings, are present. The leading edges were fitted with piano hinges so they could be opened to access systems inside; the inner section of leading edge is open to the far left. *National Museum of the United States Air Force*

The Marauder was powered by the Pratt & Whitney R-2800 Double Wasp. The engine, which depending upon model generated 1,500 to 2,800 horsepower, was the powerplant of choice for many of America's World War II combat aircraft, both fighters and bombers. The models used in the B-26 were rated from 1,850 horsepower up to 2,000 horsepower. *National Museum of the United States Air Force*

Martin engineers used this test rig during development of the B-26. In this July 24, 1940, image, exhaust back pressure on the R-2800-5 Double Wasp engine is being studied. *National Museum of the United States Air Force*

On the same day as above, this photo was taken to illustrate the external carburetor air ducts, restricted oil cooler entrance, landing-gear doors, and mockup landing gear. *National Museum of the United States Air Force*

A 1940-model Martin B-26-MA Marauder, Army Air Corps serial number 40-1464, survives in the collection of the Fantasy of Flight Museum, Polk City, Florida. During a flight to Alaska while serving with the 77th Bombardment Squadron, it crash-landed in British Columbia on January 15, 1942. The plane was recovered in 1971 and was fully restored to flying condition under the guidance of the owner of the Fantasy of Flight Museum, Kermit Weeks. *Rich Kolasa*

Specifications						
	B-26	**B-26A**	**B-26B**	**B-26C**	**B-26F**	**B-26G**
Armament:	2 x .30-cal. + 3 x .50-cal.	2 x .30-cal. + 3 x .50-cal.	11 x .50-cal.	11 x .50-cal.	11 x .50-cal.	11 x .50-cal.
Bomb load:	5,800 lbs.	5,800 lbs.	8,000 lbs.	8,000 lbs.	8,000 lbs.	8,000 lbs.
Engines (2 x):	Pratt & Whitney R-2800-5 Double Wasp	Pratt & Whitney R-2800-5 Double Wasp	Pratt & Whitney R-2800-43 Double Wasp	Pratt & Whitney R-2800-43 Double Wasp	Pratt & Whitney R-2800-43 Double Wasp	Pratt & Whitney R-2800-43 Double Wasp
Max. speed:	315 mph @ 15,000 ft.	313 mph @ 15,000 ft.	282 mph @ 15,000 ft.	282 mph @ 15,000 ft.	277 mph @ 15,000 ft.	277 mph @ 15,000 ft.
Service ceiling:	25,000 ft.	27,000 ft.	21,700 ft.	21,700 ft.	21,700 ft.	21,700 ft.
Range:	1,000 miles	1,000 miles	1,150 miles	1,150 miles	1,150 miles	1,150 miles
Wing Span:	65 ft.	65 ft.	71 ft.	71 ft.	71 ft.	71 ft.
Length:	56 ft.	58 ft., 3 in.	58 ft., 3 in.	58 ft., 3 in.	56 ft., 1 in.	56 ft., 1 in.
Height:	19 ft., 10 in.	19 ft., 10 in.	21 ft., 6 in.	21 ft., 6 in.	21 ft., 6 in.	21 ft., 6 in.
Weight (lbs.):	32,025 gross	33,022 max.	37,000 max.	37,000 max. takeoff	33,500 max. takeoff	38,200 max.
Number built:	201	139	1,883	1,210	300	893

The Fantasy of Flight Museum's B-26-MA is painted in the original camouflage scheme of Olive Drab over Neutral Gray. Propeller spinners were fixtures on the very early B-26s but were omitted starting with B-26B production. *Rich Kolasa*

Martin B-26-MA, serial number 40-1464, presents its front profile while displayed outdoors at the Fantasy of Flight Museum. Excepting the protrusion caused by the cockpit windscreen and canopy, the fuselage has a circular cross section. *Rich Kolasa*

The pronounced dihedral of the horizontal stabilizers and elevators with reference to the minimal dihedral of the wings is obvious in this photo of B-26MA, serial number 40-1464. *Rich Kolasa*

B-26MA, serial number 40-1464, is viewed from above the nose in an indoor display at the Fantasy of Fight Museum. The clear Plexiglas nose of the B-26-MAs was not reinforced, and the method of mounting the socket mount for the flexible machine gun directly on the Plexiglas would prove unsatisfactory and would be revised on the B-26B. *Rich Kolasa*

Details of the right nacelle, including the shape of the fairing for the exhaust and the panel with eight oblong vent openings on top of the nacelle, are in view. Behind a clear fairing on the leading edge of the wing is the right landing light. *Rich Kolasa*

The Plexiglas nose with its socket mount for a machine gun is viewed from above, as are the cockpit windscreen and canopy. To the front of the leading edge of the wing is the navigator's window; a similar window for the radio operator is on the opposite side of the fuselage. *Rich Kolasa*

The cockpit canopy, as seen from the upper right, features two overhead clamshell doors fitted with piano hinges on their outboard sides. Two lateral reinforcing ribs made of Plexiglas were attached to the inside of each of the clamshell doors. *Rich Kolasa*

The structure of the cockpit canopy is viewed from above and to the front, including the two clamshell doors. The pilot's and copilot's seats are visible through the windscreen. *Rich Kolasa*

The Martin Amplidyne Model 250C dorsal turret with its two Browning M2 .50-caliber machine guns traversed to the right is observed from the rear of the cockpit canopy, facing aft. Farther aft are the base of the vertical fin and the inner parts of the horizontal stabilizers. *Rich Kolasa*

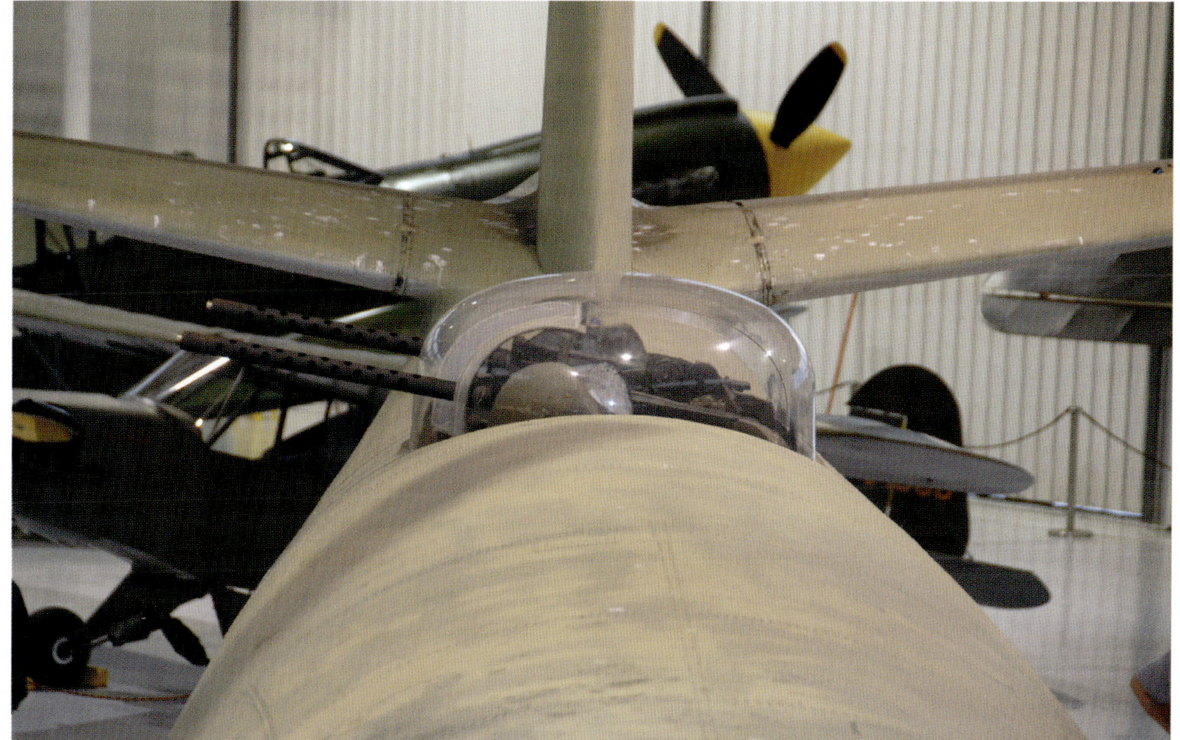

The Martin Amplidyne turret is viewed from a different angle. The clear dome was fabricated from Plexiglas. Note the perforated cooling jackets that cover the gun barrels. *Rich Kolasa*

The tail and rear fuselage of the B-26-MA are seen from the right side. The plane's current civil registration number, N4297J, is stenciled in black below the horizontal stabilizer. On the tail is the early-type tail turret. *Rich Kolasa*

The early tail turret consisted of a streamlined canopy in the tail cone of the fuselage, with a socket for a single, manually controlled .30-caliber machine gun on the top center of the canopy. *Rich Kolasa*

An observation window, curved to match the contours of the fuselage, is below the tail turret. *Rich Kolasa*

The area on the underside of the B-26-MA from the tail turret to the bomb bay is depicted. The dark-colored rectangle on the belly to the front of the tail turret is the rubber part of the tail bumper, to protect the tail during nose-high landings. *Rich Kolasa*

On the left side of the cockpit canopy, as on the right side, to the rear of the windscreen is a sliding panel. Both the windscreen and the sliding side panels were made of shatterproof glass. The rest of the canopy was of Plexiglas. *Rich Kolasa*

The bombardier's nose is constructed of two half-dome Plexiglas moldings, joined together along the horizontal center of the nose. The machine gun socket is on the upper part, and a flat panel for sighting the bomb sight is on the lower part. To the rear of that flat panel, a small door is built into the Plexiglas, so that the bombardier could reach out and wipe the flat panel if necessary. *Rich Kolasa*

The nose landing gear is viewed facing aft. The wheel was made by Bendix. Above the tire is the torque link for the oleo strut. In the top of the nose landing-gear bay, but not visible in this photo, is a hatch for accessing the forward part of the fuselage interior. *Rich Kolasa*

The left main landing gear and its bay, in the aft part of the nacelle, are viewed facing aft. The main-gear struts were hydraulically powered and of a bifolding design, wherein the bottom part of the strut pivoted forward and the upper part of the strut pivoted to the rear during retraction. *Rich Kolasa*

The right propeller, spinner, engine, and nacelle are depicted. Yellow stencils on the cuffs of the propellers provide drawing numbers, serial numbers, and data on pitch angles. On the top front of the cowling are two carburetor air intakes. On the chin of the cowling is the oil-cooler air intake. *Rich Kolasa*

Details of the manner of fabrication of the propeller spinner of the B-26-MA are shown. To the lower left are cowl flaps. *Rich Kolasa*

This Marauder is one of 109 B-26A-1-MAs built, Army serial number 41-7462. This production block incorporated one major improvement to the B-26A-MA, of which only thirty were built: the substitution of the Pratt & Whitney R-2800-39 engine for the R-2800-5. The B-26A-MA and the B-26A-1-MA had provisions for a belly rack for a torpedo, and this plane is fitted with a mockup torpedo. *National Museum of the United States Air Force*

CHAPTER 2

B-26B and B-26C

The Marauder design continued to evolve with the introduction of the B-26B in May 1942. This would become the most produced version of the Marauder. At the outset, the bombers were longer, gaining just over 2 feet through the incorporation of a redesigned tail gun installation. The new design featured two .50-caliber machine guns instead of the single weapon previously fitted. However, no seat was provided for the unfortunate rear gunner, who was obliged to kneel to operate the weapons.

A further increase in armament was made at the other end of the airplane, through the addition of two fixed .50-caliber machine guns on each side of the fuselage, below and slightly aft of the cockpit. Beneath the aircraft, more armament was added, first in the form of a tunnel-mounted .30-caliber machine gun, which was later superseded by a pair of low-mounted waist gun installations.

All these changes came with a cost in weight, and in an effort to offset this, more-powerful engines were installed beginning very early in B-26B production. In order to shorten the takeoff roll and improve performance, beginning with the B-26B-4 block the nosewheel strut was lengthened. This change increased the wing's angle of incidence. With the next production block, B-26B-5, slotted flaps began to be installed.

While in an effort to improve the combat capabilities of the Marauder, the B-26B was equipped with self-sealing fuel tanks, lessons from Europe indicated that was not enough. Thus, starting in July 1942, the Omaha Modification Center began processing 207 aircraft. Modifications included the installation of a fixed .50-caliber machine gun firing through the lower right of the nose, and the installation of a flexible weapon of the same type in the reinforced nose cone.

One of the most noticeable changes to the Marauder came with the beginning of the B-26B-10-MA production in January 1943, which was the 642nd B-26B. This and subsequent aircraft had an extended wingspan of 71 feet. This was done to lessen wing loading and improve performance and safety. At the same time, a taller fin and rudder were added, increasing the bomber's overall height from 19 feet 10 inches to 21 feet 6 inches.

With the world situation as it was in the late 1930s, many in government felt it was inevitable that the US would be drawn into the war in one capacity or another. Hence, construction began on what would be Government Owned, Contractor Operated (GOCO) facilities around the nation, many for aircraft production in the central states. One of these plants was in Omaha, Nebraska, and on January 1, 1942, that plant was leased to Martin for Marauder production. The aircraft produced there, essentially duplicates of the Baltimore-built B-26B with extended wingspan, were designated B-26C. Production began in Omaha in August 1942.

A power-assisted tail gun turret was introduced with production block B-26C-20-MO. The Bell-designed turret provided a greater field of fire than that of the manual installation and allowed the airframe to be shortened by 2 feet. A similar change was made on the Maryland production line as well.

The Omaha plant also turned out 123 B-26Cs for the British Commonwealth, which designated them Marauder II. The aircraft, which bore RAF serial numbers FB400 through FB522, were ordered for Lend-Lease by the US Army under contract AC-19342. The Commonwealth subsequently assigned the bulk of the aircraft to the South African Air Force.

Similar to the B-26C, the AT-23B target tug was also produced in Omaha. The Army, upon request, transferred 225 AT-23B aircraft to the Navy, which designated them JM-1. Marauder production was halted in Omaha in April 1944, in order to allow the plant to be converted for B-29 Superfortress production.

The B-26B-MA introduced a redesigned tail stinger with twin .50-caliber machine guns and increased ammunition capacity. Spinners now were omitted from the propellers, Pratt & Whitney R-2800-41 engines were installed, and the aircraft's electrical system was changed from 12 volt to 24 volt. Here, B-26B-MAs are lined up at the Middle River plant, with shipment stalled because of a shortage of Curtiss Electric propellers. *National Museum of the United States Air Force*

Martin B-26B-MA, army serial number 41-17627, is fitted with propeller spinners, a feature that would be deleted from the Marauders during B-26B-MA production. The short wingspan of this model and previous B-26 models resulted in overly fast takeoff and landing speeds, and the stubby vertical tail led to stability problems. This plane had the short wingspan and low tail, but these problems would be addressed later during B-26B and B-26C production. *National Museum of the United States Air Force*

Martin B-26B-2-MA, US Army Air Forces (as the Army Air Corps had been renamed) serial number 41-17876, is the subject of the following sequence of in-flight photos. Although this plane is unarmed, the B-26B-2-MAs had provisions for a fixed machine gun in the lower right corner of the clear nose. The upper parts of the engine nacelles had been redesigned and now featured expanded carburetor intakes. *National Museum of the United States Air Force*

A centerline torpedo rack is visible on the belly of B-26B-2-MA, serial number 41-17876. To the front of the rack is the teardrop-shaped radio direction-finder (RDF) antenna housing, sometimes called the football antenna. The small window in the fuselage to the front of the wing root was next to the radio operator's station. *National Museum of the United States Air Force*

The tail stinger is visible in this photograph taken from the left rear quarter of B-26B-2-MA, serial number 41-17876. The machine guns are not mounted in this view, but they protruded through an opening at the rear of the canopy, above the tail cone. This stinger is sometimes described as having a "stepped" design. The national insignia is the type without the red circle, which was authorized from May 1942 to June 1943. *National Museum of the United States Air Force*

The control surfaces—the rudder, the elevators, and the ailerons—as seen on B-26B-2-MA, serial number 41-17876, are a darker color than the adjacent skin, due to the different materials, fabric on the control surfaces versus aluminum alloy on the skin, and the different manners of paint application and absorption. On top of the fuselage between the cockpit canopy and the fronts of the wings is a barely discernible round escape hatch. *National Museum of the United States Air Force*

As seen in a view of B-26B-2-MA, serial number 41-17876, from below, a landing light was in the leading edge of each wing, behind a clear cover. A good view is available of the wavy demarcation between the upper Olive Drab paint and the lower Neutral Gray. The spinners are no longer present on the propellers. *National Museum of the United States Air Force*

This B-26B has the short-type wings used up to the B-26B-10-MA and B-26C-5 production blocks. A distinguishing feature of the short wings as seen from this angle was the aileron trim tabs, which extended slightly to the rear of the trailing edges of the ailerons. On the long wings, the rears of the trim tabs were flush with the rears of the ailerons. *National Archives*

In a final photo of B-26B-2-MA, serial number 41-17876, in flight, the pronounced dihedral of the horizontal stabilizers and elevators, 8 degrees, is apparent. Also apparent are the sleek, streamlined lines of the B-26, which gave it an image of bullet-like speed compared to some of the less graceful US bombers of the early World War II era. *National Museum of the United States Air Force*

B-26B-MA, serial number 41-17704, exhibits a tail stinger with a tail cone below and to the rear of it. The aircraft's number, 75, was marked on the dorsal fin and the side of the nose. This plane was written off after an accident near Pensacola, Florida, on November 26, 1942. *National Archives*

Sporting shark's-mouth and shark's-eyes artwork, Martin B-26B-45-MA, serial number 42-95793, served with the 444th Bombardment Squadron, 320th Bombardment Group. The navigator's and radio operator's escape hatch, atop the fuselage to the front of the wings, is highly visible in this view. The plane was written off as damaged beyond repair on May 11, 1944. *National Archives*

B-26B-45-MA, serial number 42-95739, runs up its engines. With production block B of the B-26B model, the aft bomb bay was converted into ammunition storage space, and the aft bomb bay doors were sealed. The clear nose features metal bracing for the .50-caliber machine gun, now mounted at the front point of the nose. Note the bulged window at the navigator's station, above the package machine guns. *National Museum of the United States Air Force*

A checkerboard ring has been painted to the immediate rear of the clear nose on B-26B-MA, serial number 41-17694. The enlarged carburetor intakes on the tops of the nacelles were introduced to the production line with the B-26B-3-MA but were retrofitted on earlier-production B-26Bs such as this plane. Like so many early B-26 Marauders, this one met an early end, being damaged beyond repair following an accident during takeoff from a base in Florida on June 2, 1943.
National Museum of the United States Air Force

Martin B-24B-MA, USAAF serial number 41-17704, is seen from a slightly different perspective during flight. The slight bump visible on the bottom of the fuselage below the horizontal stabilizer is a skid, to protect the skin and structure of the empennage during nose-high landings or takeoffs. *National Museum of the United States Air Force*

Faintly visible on the side of the fuselage adjacent to the cockpit of B-26B-55-MA, serial number 42-96123, are two panels of appliqué armor; these were referred to as pilot's ground-strafing armor. This Marauder features the Bell M6 electrohydraulically powered tail turret, which was introduced to Marauder production with the B-26B-20-MA and B-26C-10-MA production blocks. On the side of the fuselage are two .50-caliber package machine guns, introduced with the B-26B-10-MA production block. *National Museum of the United States Air Force*

This B-26 is a real curiosity. Only part of the tail number is visible, so the exact model and production run are not known. The cowl with its low-profile carburetor air intakes appears to be of a B-26B-2-MA or earlier. The clear nose has been braced for a .50-caliber machine gun at the tip of the nose. Very unusual is the cheek window to the rear of the clear nose. The rudder is an experimental type with a forward-pointing counterbalance at the top. Finally, the plane sports a Bell M6–powered tail turret. *National Museum of the United States Air Force*

The New York Central Railroad purchased this B-26B-2-MA, USAAF serial number 41-17916, for the Army Air Forces. A commemorative inscription to that effect is on the side of the fuselage, aft of the clear nose. In the background is a line of B-26 Marauders. *National Museum of the United States Air Force*

Officials conduct an unveiling ceremony for B-26B-2-MA, serial number 41-17916, donated to the Army Air Forces by the New York Central Railroad. Patriotic bunting partially covers the inscription, which reads, "THE / NEW YORK / CENTRAL / SYSTEM / GIFT OF THE EMPLOYEES / OF THIS RAILROAD." *National Museum of the United States Air Force*

A security guard stands watch over newly completed B-26Bs. The nearest Marauder is B-26B-MA, USAAF serial number 41-17839. The tail stinger is the type with the tail cone to the lower rear of the canopy. Note the small window to the front of the national insignia. On the front of the wing is a deicer boot, a rubber assembly that was slightly inflated to break up ice buildup. *National Museum of the United States Air Force*

The gunner is at his station in the tail stinger of B-26B-10-MA, serial number 41-18319, assigned to the 437th Bombardment Squadron, 319th Bombardment Group. The twin Browning M2 .50-caliber machine guns protruded through a square opening in the canopy, and the gunner operated the guns manually, with his hands on the grips. *National Museum of the United States Air Force*

A shot-up Bell M6 tail turret is viewed from the right rear in a photo dating from late November 1943. The one-piece Plexiglas canopy is missing, showing the ammunition chutes and the N8 gunsight and mount to the rear of the gunner's armor plate, with a ballistic-glass sighting window. This turret in many ways was an improvement over the previous tail stinger, since it featured electrohydraulically powered gun traverse and elevation and had a wider field of fire. *National Museum of the United States Air Force*

B-26B-20-MA, serial number 41-31746, fuselage code ER-K, of the 450th Bombardment Squadron, 322nd Bombardment Group, is seen after being badly shot up. This was likely the same aircraft seen in the preceding photo. Between the "ER" code and the national insignia is the right waist window, which accommodated a .50-caliber machine gun. Waist machine guns that fired through small windows had been present since the B-26B-1-MA, but after combat experience in Europe, the waist windows were enlarged and moved one station aft commencing with the B-26B-10-MA. *National Museum of the United States Air Force*

As a ground crewman refuels a four-engine aircraft, B-26s are lined up in the background at an airfield at Rufisque, French West Africa (now Senegal), in June 1943. The nearest Marauder is "Lady Katy," B-26B-10-MA, USAAF serial number 41-18285. The three planes beyond "Lady Katy" are B-26C-5-MOs. Where the rears of the fuselages are visible, all the B-26s have the stepped-type tail stingers. *National Museum of the United States Air Force*

Bombs are ready for loading into Marauders next to B-26B-55-MA, serial number 42-96205, fuselage code X2-S, serving with the 596th Bombardment Squadron, 397th Bombardment Group, 9th Air Force. The diagonal stripe on the tail was yellow with black borders. Invasion stripes are painted on the wings and belly, which was a recognition device employed at the time of the June 6, 1944, landings in Normandy, and for some time thereafter. *National Museum of the United States Air Force*

Marauders of the 449th Bombardment Squadron, 322nd Bombardment Group, 8th Air Force, swoop in during a bombing mission as flak bursts pepper the sky in the distance. To the left is B-26B-10-MA, serial number 41-18272 (PN-Q), and in the background is B-26C-20-MA, serial number 41-34683 (PN-V). This photo probably was taken around May or June 1943; the 332nd Bomb Group flew its first missions in May 1943, and the national insignia seen on the planes was discontinued in June 1943. *National Museum of the United States Air Force*

The British acquired a version of the Martin B-26 based on B-26A-MA and B-26A-1-MAs, dubbing it the Marauder I. In the foreground is Marauder I RAF serial number FK375, nicknamed "Dominion Revenge," based on a B-26A-1-MA airframe and assigned to No. 14 Squadron, based in Egypt in 1942 and 1943. Two other Marauders are in the background. *National Museum of the United States Air Force*

"Shootin' In," B-26B-50-MA, serial number 42-95857, FW-K, of the 556th Bombardment Squadron, 387th Bombardment Group, drops a load of bombs on an enemy target. Although much smaller than the B-17 Flying Fortress, the B-26s were capable of carrying approximately the same capacity of bombs: a maximum of 8,000 pounds. The B-17, of course, had the advantage in range and service ceiling. *National Museum of the United States Air Force*

A Marauder in bare-aluminum finish, B-26B-55-MA, serial number 42-96200, releases its bomb load over enemy territory. The fuselage of this plane was RJ-M, and it served with the 323rd Bombardment Squadron, 454th Bombardment Group. Note the open waist window, with a view through the open left waist window. *National Museum of the United States Air Force*

A mix of models of Marauders with shark's-mouth artwork, serving with the 444th Bombardment Squadron, 320th Bombardment Group, salvo their bombs. From left to right, they are B-26B-10-MA, serial number 41-18260; B-26B-50-MA, serial number 42-96013; B-26B-40-MA, serial number 42-43277; and B-26B-50-MA, serial number 42-95946. The plane to the left has the stepped-type tail stinger, while the others have the Bell-powered turrets. *National Museum of the United States Air Force*

A heavily weathered B-26B-55-MA assigned to the 599th Bombardment Group, 397th Bombardment Group, USAAF serial number 42-96165, and fuselage code 6B-I, displays a shark's-mouth design and invasion stripes. Note how the stripes were cut in just short of the national insignia on the bottom of the right wing. *National Archives*

In a war in which airplanes and aircrews did not enjoy good odds for surviving many missions, "Mild and Bitter," B-26B-25-MA, serial number 41-31819, DR-X, beat the odds by becoming the first Marauder to complete fifty missions. Named after a British beer mixture, "Mild and Bitter" was assigned to the 452nd Bombardment Squadron, 322nd Bombardment Group, 9th Air Force, at Great Saling (Station 485) in Essex, England. The plane went on to complete 100 missions, after which it was sent back to the States to participate in a war bond drive. *National Museum of the United States Air Force*

"Flak Bait" undoubtedly was the most famous of the Martin Marauders. The plane, B-26B-25-MA, serial number 41-31773, was assigned to the 449th Bombardment Squadron, 322nd Bombardment Group, and flew 207 operational missions over Europe, earning it the status of the USAAF plane that completed the most missions during World War II. "Flak Bait" was preserved after returning to the United States, and it is currently in the collection of the National Air and Space Museum. *National Museum of the United States Air Force*

Firefighters struggle to put out the flames after B-26B-55-MA, USAAF serial number 42-96093 and fuselage code X2-V, crashed during a transition-training flight at Rivenhall (Station 169), England, on July 28, 1944. The plane was assigned to the 596th Bombardment Squadron, 397th Bombardment Group. Almost two months after D-day, the plane had still been wearing invasion stripes. *National Museum of the United States Air Force*

"You Cawn't Miss It" was the nickname of this B-26B assigned to the 494th Bombardment Squadron, 344th Bombardment Group. The name of the pilot, J. R. Ashberry, is painted below the sliding side panel of the cockpit canopy. The cover of the top package machine gun is open. *Roger Freeman collection*

B-26B-50-MA, serial number 42-95952, was nicknamed "You've 'ad It." It was assigned to the 497th Bombardment Squadron, 344th Bombardment Group. Admiring the nose art is Lt. Jack Havener, of the 344th Bombardment Group. *Roger Freeman collection*

"Hard to Get," B-26B-50-MA, serial number 42-95903, featured on the right side of the forward fuselage a pinup art of a reclining beauty. This Marauder was assigned to the 344th Bombardment Group. *Roger Freeman collection*

To meet USAAF demands for aircraft, Martin entered production at a second plant, at Bellevue, Nebraska, a suburb of Omaha, manufacturing the B-26C series. This plant generally was referred to as the Omaha plant. From the start, the B-26Cs were equipped with the long-span wings and the enlarged vertical tail. The first production block was the B-26C-5-MO, and shown here is the sixth example to leave the production line, USAAF serial number 41-34678. *National Museum of the United States Air Force*

The B-26C-5-MO had the lengthened nose landing gear that also began appearing on the B-26B with production block 4. This was characterized by a bulge that was split between the chin of the fuselage and the fronts of the nose landing-gear doors. On the B-26B the crew of the front portion of the plane entered it through the nose landing-gear bay; starting with the B-26C-6-MO, a crew door was installed on the lower right side of the fuselage, to the front of the bomb bay door. *National Museum of the United States Air Force*

A stepped-type tail stinger was mounted on the B-26C-5-MO. In this example, twin .50-caliber machine guns have been mounted. A skid was on the bottom of the fuselage, and the small-type waist windows were present. *National Museum of the United States Air Force*

President Franklin D. Roosevelt visited the Martin plant near Omaha on April 26, 1943. He is seen here in the front seat of the limousine, with Glenn L. Martin (in the dark suit) and Nebraska governor Dwight Griswold in the rear seat. Here, they are reviewing newly manufactured Marauders; to the left is B-26C-20-MO, serial number 41-35201. *National Museum of the United States Air Force*

Martin B-26C-20-MO, serial number 41-34685, sports an extra probe on top of the left pitot tube, similar to the one that had been installed on the first B-26-MA. Protruding through the open waist window is a .50-caliber machine gun. The machine gun barrels are light colored because they had a protective wrapping. Note the two small, round windows above the waist window and the enlarged carburetor air intakes on top of the cowl. *National Museum of the United States Air Force*

Martin Marauders of the 386th Bombardment Group are lined up on the margins of a taxiway at Bassingbourne (Station 121), England, around early November 1943. These planes, from what can be discerned of their tail numbers, appear to be a mix of B-26Bs and B-26Cs. The nearest plane at the center of the photo has the RU fuselage code of the 554th Bomb Squadron; a plane to the right has the RG code of the 552nd Bomb Squadron. *National Museum of the United States Air Force*

The bomb bay doors are open on B-26C-25-MO, USAAF serial number 41-35358, fuselage code YA-V, as it prepares to release a load of bombs. This plane was attached to the 555th Bombardment Squadron, 386th Bombardment Group. As early as August 1943, this plane had been going by the nickname "Sexy Betsy," with nose art on the right side of the fuselage. At that time, the aircraft was based at Great Dunmew, Essex, England. *National Museum of the United States Air Force*

"The Yankee Guerrilla" was the nickname of B-26C-20-MO, serial number 41-34946, of the 555th Bombardment Squadron, 386th Bombardment Group. Although not visible here, the plane's fuselage code was YA-L. The nose art is of a top-hatted figure wielding a bomb in each hand. The name "GAIL" is marked on the left engine cowl. *National Museum of the United States Air Force*

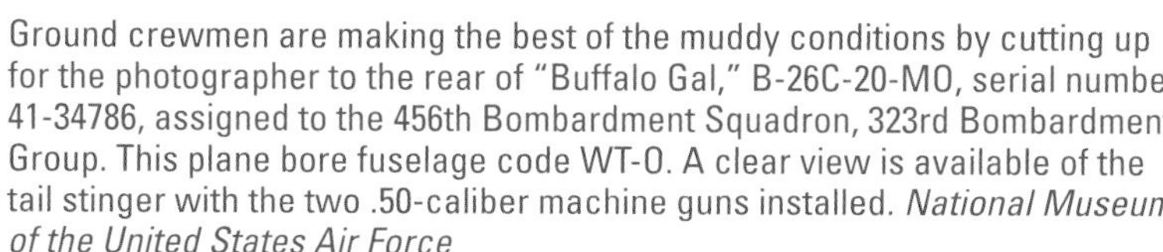
Ground crewmen are making the best of the muddy conditions by cutting up for the photographer to the rear of "Buffalo Gal," B-26C-20-MO, serial number 41-34786, assigned to the 456th Bombardment Squadron, 323rd Bombardment Group. This plane bore fuselage code WT-O. A clear view is available of the tail stinger with the two .50-caliber machine guns installed. *National Museum of the United States Air Force*

A B-26C-45-MO, USAAF serial number 42-107798, of the 95th Bombardment Squadron, 17th Bombardment Group, 1st Tactical Air Force, darts over a destroyed rail yard in a portion of Florence, Italy, still held by the Germans, in order to perform an assessment of bomb damage to the facility. A large number "60" is painted on the vertical tail. *National Museum of the United States Air Force*

Martin B-26C-45-MO, serial number 42-107798, of the 95th Bombardment Squadron, 17th Bombardment Group, 1st Tactical Air Force, flies over the edge of a city, possibly Florence. The tail turret is the Bell M6A, on which the twin .50-caliber machine guns protruded through a canvas boot instead of a Plexiglas dome, as on the M6 turret. The protrusion below the turret is a spent-casing collector. *National Museum of the United States Air Force*

Two Marauders of the 454th Bombardment Squadron, 323rd Bombardment Group, 9th Air Force, fly in close formation during a bombing mission. At the top is B-26C-45-MO, serial number 42-107614, RJ-H, below which is B-26C-25-MO, serial number 41-35253, RJ-S. Interestingly, the white invasion stripes on both planes have been overpainted with a dark color that shows as slightly lighter than the remaining black stripes. *National Museum of the United States Air Force*

Martin B-26 Marauders of the 454th Bombardment Squadron, 323rd Bombardment Group, 9th Air Force, return from an air strike in support of the 3rd US Army along the German-French border on November 16, 1944. These include the two planes seen in the preceding photo: B-26C-25-MO, serial number 41-35253, RJ-S (in the left foreground), and B-26C-45-MO, serial number 42-107614, RJ-H (to the right). *National Museum of the United States Air Force*

"Classy Chassy" was B-26C-45-MO, USAAF serial number 42-107665 and fuselage code PN-X. The plane served with the 449th Bombardment Squadron, 322nd Bombardment Group, 9th Air Force. The faint outline of a national insignia with side bars is faintly visible between the national insignia and the X. *National Museum of the United States Air Force*

On May 20, 1944, during a mission over Normandy, this B-26C-45-MO piloted by Lt. Col. John S. Samuel, deputy commander of the 391st Bombardment Group, was hit by flak, badly shredding the tail. Despite the destruction, Samuel was able to bring the plane back to base and land it by carefully regulating the engines to steer the aircraft. This Marauder, USAAF serial number 42-107740 and fuselage code T6-U, was assigned to the 573rd Bombardment Squadron. *National Museum of the United States Air Force*

A member of the crew of B-26C-45-MO, USAAF serial number 42-107582, poses next to the plane. This Marauder was assigned to the 454th Bombardment Squadron, 323rd Bombardment Group, based at Earls Colne (Station 358), England. It bore the fuselage code RJ-B; subsequently, it was assigned the code RJ-R. Note the reinforcing spline on the leading edge of the wing, near the wing root. *National Museum of the United States Air Force*

The Free French acquired a number of B-26s and organized them into several squadrons. This plane, B-26C-45-MO, serial number 42-107657, was assigned to *Groupe de Bombardement Moyen* (GBM, Medium Bombardment Group) 1/22 "Maroc," of *31 Escadre* (Squadron), attached to the 42nd Wing of the 12th US Army Air Force. *National Archives*

Martin B-26C-45-MO, serial number 42-107827, serving with the French air force, flies a bombing mission in the area around Boulogne, France. A shield-type insignia is on the fuselage, alongside the cockpit. Note the different types of national insignia, several with bars and one without bars. A large number "55" is on the tail. *National Museum of the United States Air Force*

A photographer captured this image of B-26C-45-MO, serial number 42-107709, returning from an airstrike against a railroad bridge in the south of France. A large number "07" is superimposed over the tail number. The plane had a bare-aluminum finish. *National Museum of the United States Air Force*

French air force B-26s return from an attack on German communications networks in northwestern Italy. To the left is the tail of B-26C-45-MO, serial number 42-107715, aircraft number 35, adjacent to which is B-26C-45-MO, serial number 42-107833 (or 42-107838), number 40. On that plane's fuselage, by the cockpit, is a cross of Lorraine, symbol of the Free French forces.
National Museum of the United States Air Force

In a photograph dated July 20, 1944, on the negative, B-26C-45-MO, serial number 42-107685, flies above the English Channel on a mission. At the time this Marauder was serving with the 450th Bomb Squadron, 322nd Bomb Group, 9th Air Force, and was wearing fuselage code ER-V. Subsequently, this plane was transferred to the 558th Bomb Squadron, 387th Bomb Group, 9th Air Force, and was shot down by antiaircraft fire on February 22, 1945; its crew managed to bail out. *National Museum of the United States Air Force*

SSgt. Richard J. Hightower of Macon, Missouri, mans a .50-caliber machine gun in the left waist window of a B-26. He was assigned to the 444th Bombardment Squadron, 320th Bombardment Group, 12th Air Force, and at the time this photo was taken he had flown on forty missions. At the bottom of the photo is the swing arm that the machine gun was mounted on. The dark structure to Hightower's rear is the lower front of the turret. To the left is a roller track carrying ammunition from the former aft bomb bay to the tail turret. *National Archives*

The left waist window and Browning M2 .50-caliber machine gun of a B-26 is depicted. The gun is mounted on a fixture called an adapter, which contained handgrips and a mount for the gunsight and served as a buffered frame to hold the gun. The adapter also had fittings for mounting it on the pintle mount. On the fuselage, to the front of the waist window, are two wind deflectors, which were deployed when the waist window was open during flight to keep air from rushing into the window. *National Museum of the United States Air Force*

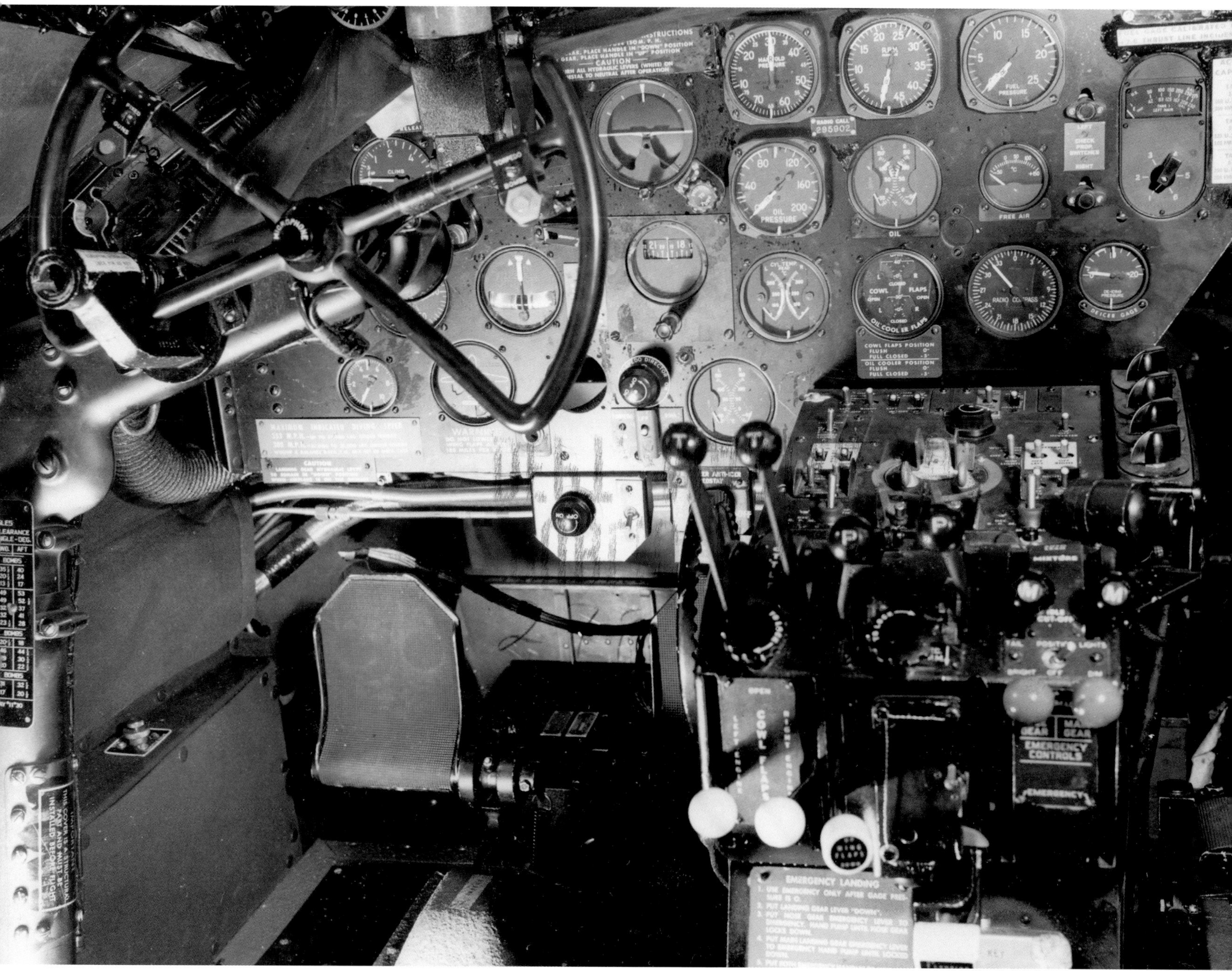

This is a pilot's-eye view of the cockpit, showing his control yoke, the instrument panel, and, to the lower right, the pedestal. To the far left is the control column on which the control yoke was mounted. The copilot had a control column and yoke that were the mirror image of the pilot's. The instrument panel did not extend all the way across the cockpit but was terminated to the right of the pedestal, allowing a passageway from the copilot's side of the cockpit to the bombardier's compartment.
National Museum of the United States Air Force

The pedestal is viewed from the rear. The hand to the left is on the control for the left engine cowl flaps, below which is a placard with instructions for emergency landing. The two tall levers with ball grips on the left side of the top of the pedestal are the left and right engine throttle controls. At the bottom of the pedestal are controls for the engine blowers, the oil cooler shutters, and the carburetor air temperature. *National Museum of the United States Air Force*

A B-26C instrument panel is shown in a photo dated December 23, 1943. Above the instrument panel are a compass and a dust cover. The oblong instrument to the far right is the fuel gauge. The instrument with the outline of the airplane on it, toward the lower left, is the landing-gear and wing-flaps indicator. *National Museum of the United States Air Force*

A pilot (*left*) and copilot are shown in a B-26C cockpit in a December 23, 1943, photograph. The large, knurled knob on the console at the top center of the photo is the aileron-tab control. The pilot had a high-backed seat, while the copilot's seat had a low back. Note the ring sight above the instrument panel on the pilot's side, for aiming the package machine guns. *National Museum of the United States Air Force*

“Sky Queen,” B-26C-25-MO, serial number 41-35264, and fuselage code TQ-U, served with the 559th Bombardment Squadron, 387th Bombardment Group, 9th Air Force, in World War II. The nose art featured a flying nude woman and twenty-seven bomb symbols indicating combat missions.

Martin B-26C-10-MO, serial number 41-34891, went by the nickname “Missouri Mule” and carried a large number “19” on the tail, while serving with the 441st Bombardment Squadron, 320th Bombardment Group. The pilot was Lt. Robert Dinwiddie. To the front of the “Missouri Mule” inscription on the nose is the insignia of the 441st Bombardment Squadron. *Roger Freeman collection*

"Missouri Mule" is observed from another angle in flight. Note the yellow band below the horizontal stabilizer. *Roger Freeman collection*

The navigator of a Marauder from the 441st Bombardment Squadron snapped this color photo of "Missouri Mule" through his window, also capturing a view of the right cowling and propeller of his aircraft. *Roger Freeman collection*

In the spring of 1944, production of B-26Cs at Martin's Omaha plant drew to an end, and the facility shifted to the production of B-29 heavy bombers. Here, the final Marauder produced at Omaha, B-26C-45-MO, serial number 42-107855, is rolled out of the assembly line. The sign on the plane reads, "Omaha [with 'B-26' inside the O] / One Victory Won / *Greater Victories* to come." The placard by the clear nose reads, "A 44/Block 45." *National Museum of the United States Air Force*

CHAPTER 3

B-26F and B-26G

Production of the next mass-produced Marauder model, the B-26F, began in Baltimore late in February 1944. The B-26F is distinguishable from the earlier aircraft via a change in the wing, which was installed with a 3.5-degree increase in angle of incidence. This change yielded a reduction in takeoff speed to 110 miles per hour and, more significantly, reduced the takeoff run for a combat-loaded Marauder by about 100 yards. The upward cant of the wing and engines also increased propeller ground clearance, an important consideration given the crude airfields from which medium bomb squadrons often operated. Unfortunately, the cost of this change was increased drag, which reduced the top speed of the Marauder by about 5 mph.

After building a scant 100 B-26Fs for the US Army Air Forces, Martin then produced a further 200 for the Royal Air Force. These aircraft, assigned RAF serial numbers HD402 through HD601, began leaving the assembly line in March 1944. Like their Commonwealth predecessors, the bulk of these aircraft went to the South African Air Force. The Commonwealth aircraft differed from USAAF aircraft in that the Commonwealth preferred the Bell M6A tail turret, with canvas closure, versus the American Bell M6, with Plexiglas closure.

Externally almost indistinguishable from a Marauder III, the final production version of the B-26 was the B-26G. Most significantly, the B-26G replaced many Army-only components with standardized Army-Navy (AN) items, particularly the radios. Externally, the B-26G could be distinguished from the earlier B-26F due to the canvas-enclosed guns of the M6A, rather than the Plexiglas cone of the M6 tail turret. Also visible was a larger life raft compartment in the top of the forward fuselage.

As B-26G production hit stride, the Marauder's armament had stabilized with an impressive eleven .50-caliber machine guns. One of these was mounted in a flexible nose position, with two in the upper turret, two in the tail turret, one in each waist position, and four in single-gun blister packs, two of which were mounted on each side of the nose.

Like the B-26F before it, the B-26G left the Middle River plant in natural-aluminum finish, rather than camouflage. All 893 B-26G models were built by Martin at the Middle River plant near Baltimore. One hundred and fifty of these were supplied to Britain under the Lend-Lease Act (RAF serial numbers HD602-HD751), where like the 200 B-26Fs they were designated Marauder III.

A substantial number of B-26G-5-MA aircraft were delivered to the Free French air force. In fact, today two of the four surviving complete B-26s are ex–Free French air force aircraft.

The final B-26G, and indeed the final B-26, rolled off the assembly line April 18, 1945. Dubbed "Tail End Charlie—"30," the bomber had at the controls William Ebel, Martin chief engineer and the man who took the very first B-26 into the air almost five years earlier.

By 1948, the Marauder had been classified as obsolete by the US military, and the wholesale disposal, often by scrapping—sometimes preceded by explosive demolition—was well under way.

The Baltimore-produced Martin B-26F, introduced in early 1944, featured wings with an angle of incidence of 3.5 degrees more than that of preceding B-26 models. This modification was designed to allow the plane to make shorter takeoffs and landings at lower speeds. Depicted here is the second B-26F-1-MA, USAAF serial number 42-96230. *National Museum of the United States Air Force*

The increased-incidence wing of the B-26F was first tested on B-26B-40-MA, serial number 42-43459, as seen here. Offsetting the landing and takeoff performance benefits of the higher-incidence wing was increased drag during flight, occasioned by the tail-up attitude the aircraft experienced with these wings. *National Museum of the United States Air Force*

Martin B-26B-40-MA, serial number 42-43459, is viewed head on during a flight test of the increased-incidence wings. The slight drag induced by the tail-up attitude of the plane during flight resulted in a somewhat reduced maximum airspeed: 277 miles per hour at 10,000 feet, as compared with the top speed of the stock B-26B-40-MA, 282 miles per hour at 15,000 feet. *National Museum of the United States Air Force*

During a test flight, B-26B-40-MA, serial number 42-43459, is observed from above. From this angle, the plane looks much like any other B-26B, except for the slightly more constricted width of the top deck of the fuselage between the forward parts of the wing roots.
National Museum of the United States Air Force

A final photo of the B-26B-40-MA test bed for the increased-incidence wings designed for the B-26F captures the plane from the lower right quarter. Noticeable features include the bulge at the front of the nose landing-gear doors, the package guns and the bulged navigator's window above them, and the wind deflectors to the front of the waist window. *National Museum of the United States Air Force*

What little is visible of the tail number of this Marauder—that is, the first four digits, 2963—identifies it as a B-26F-1-MA or B-26F-2-MA. The absence of a full tail number and the partially obscured fuselage code make it difficult to confirm the unit of this plane, but some sources have identified it as code K5-U of the 584th Bombardment Squadron, 394th Bombardment Group, and the diagonal stripe on the vertical tail is consistent with that group's markings in the final months of the war. *National Museum of the United States Air Force*

During 1944, the United States transferred one hundred B-26F-2-MAs, one hundred B-26F-6-MAs, seventy-five B-26G-11-MAs, and seventy-five B-26G-21-MAs to the Royal Air Force and South African Air Force, who designated them as Marauder IIIs. Shown here is Marauder III, serial number HD561, of No. 21 Squadron, South African Air Force. The scoreboard of this plane shows bombs representing 113 missions. On the nose is the letter "B" and cartoon art of a black cat with an illegible word underneath. *National Museum of the United States Air Force*

The sixth B-26F-1-MA, USAAF serial number 42-96234, was experimentally fitted with a Hamilton Standard reverse-pitch propeller and is seen here at an airport using its engines and propellers to back up. The top turret had been removed and replaced with a form-fitting clear panel with a cross-shaped frame. *National Museum of the United States Air Force*

Martin B-26F-1-MA, serial number 42-96322, assigned to the 441st Bombardment Squadron, 320th Bombardment Wing, 12th Air Force, was photographed from another B-26 during a mission in the latter part of World War II. This was one of a relatively few B-26Fs to serve with the 12th Air Force in the Mediterranean theater. *Roger Freeman collection*

The B-26G introduced a number of improvements to the B-26F, including a transition to standardized Army-Navy interior equipment. The mast over the bomb bay, introduced during B-26F production, was not for a radio antenna but instead was a vent for the fuel tanks. The B-26G also marked a return to a pitot tube on the wing, but on the left wing only. Seen here is the sixth B-26G-1-MA, USAAF serial number 43-3496. *Air Force Historical Research Agency*

A B-26G appears in a photo taken in mid-August 1944. Note the fuel vent to the rear of the radio antenna mast, the pilot's appliqué ground-strafing armor by the cockpit, the nomenclature and data stencil to the front of the armor, and the reinforcing spline on the leading edge of the wing near the wing root, which was a feature added to Marauders in the early stages of production. *National Museum of the United States Air Force*

A B-26G runs up its engines on a hardstand. Several nacelle panels are removed, indicating that the plane was likely undergoing diagnostic tests. This plane, like many B-26Gs, was painted in a partial camouflage of Olive Drab on the upper surfaces and bare aluminum on the lower ones. *National Museum of the United States Air Force*

Martin B-26G-1-MA, serial number 43-34132, appears in a highly polished bare-aluminum finish with no markings other than the national insignia and the tail number. This early-production B-26G-1-MA still has the pitot tube on the bottom of the fuselage to the front of the nose gear; later, this would be deleted and the pitot tube would be located on the left wing. *National Museum of the United States Air Force*

The B-26Gs such as B-26G-15-MA, serial number 44-67824, shown here, were armed with the Bell M6A tail turret. On this turret, the Plexiglas dome through which the .50-caliber gun barrels protruded was replaced with a simple boot made of fabric. A spent-casing collector is on the bottom of the fuselage, below the tail turret. *National Museum of the United States Air Force*

Only the first four digits of the tail number of this Marauder are visible, 3343, but it is enough to identify this as a B-26G-5-MA. The B-26Gs had a large, rectangular hatch to the rear of the pilot's and copilot's escape hatches, with two small, round windows in it. To the rear of that hatch was the round escape hatch that had long been on the B-26s. Note how the fuel vent was offset to the right of center. *National Museum of the United States Air Force*

B-26G-5-MA, serial number 43-34396, cruises at high altitude off a coastline. To the far right is the pitot tube, which had been relocated to the left wing. The clear landing-light housing on the left wing of the B-26 was wider than that on the right wing because the left housing enclosed a red passing light in addition to the landing light, while the right wing had only the landing light. *National Museum of the United States Air Force*

A serviceman stands by a pile of bombs ready for loading into bombers, with B-26G-5-MA, serial number 43-34342, called "Lil Maggie," in the background. This plane was assigned to the 322nd Bombardment Group and wore a partial camouflage scheme of Olive Drab over bare aluminum.
National Museum of the United States Air Force

Martin B-26G-1-MA, serial number 43-34194, fuselage code K5-S, was assigned to the 584th Bombardment Squadron, 394th Bombardment Group. The plane had a partial camouflage scheme, with the Olive Drab flaking off heavily around the top turret and the trailing edge of the rudder and trim tab. Note the dark-colored boot on the rear of the Bell M6A turret. *National Museum of the United States Air Force*

"Porky," B-26G-5-MA, USAAF serial number 43-34287 and fuselage code TQ-X, is seen following a crash landing occasioned by landing when out of fuel, at Advanced Landing Ground A-62D, Reims, France, on February 9, 1945. The plane served with the 559th Bombardment Squadron, 387th Bombardment Group, and was based at Advanced Landing Ground A-71 at Clastres, France, at the time of the crash. The plane was written off as unrepairable. On the tail is the diagonal-striped marking of the 387th Bomb Group. *National Museum of the United States Air Force*

Engine maintenance is underway on "La Paloma" following a snowstorm. This plane was B-26G-1-MA, USAAF serial number 43-34210 and fuselage code AN-V, assigned to the 553rd Bombardment Squadron, 386th Bombardment Group, 9th Air Force, based at Beaumont-sur-Oise Airfield (A-60), France, in the winter of 1944–45. *National Museum of the United States Air Force*

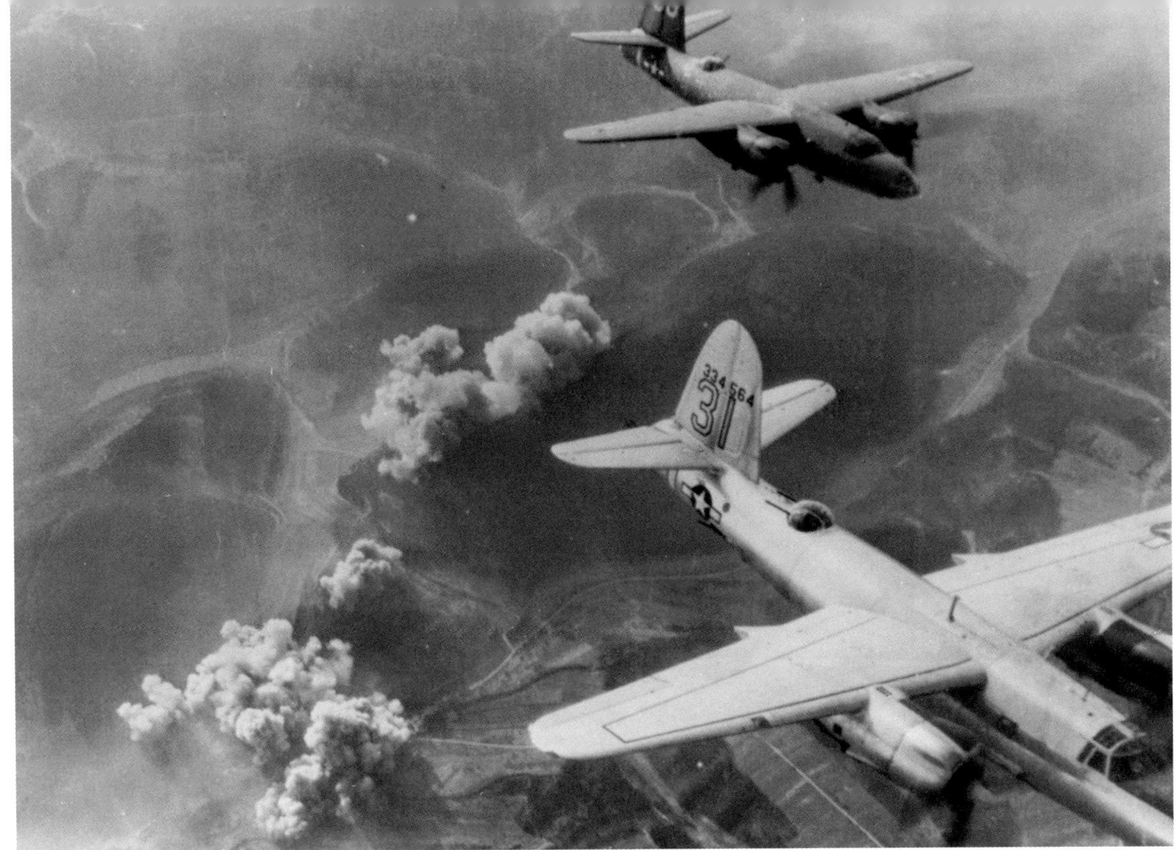

Marauders of the 1st Tactical Air Force bomb enemy positions along the Siegfried Line near Erlenbach, Germany, in support of 9th US Army operations. In the foreground is B-26G-11-MA, serial number 43-34564 and aircraft number 31, while in the background is a B-26 numbered 33. *National Museum of the United States Air Force*

The enhanced angle of incidence of the wing of this half-camouflaged B-26G-25-MA is particularly noticeable in this photograph. This Marauder, USAAF serial number 44-68118 and fuselage code 4T-E, flew with the 585th Bombardment Squadron, 394th Bombardment Group. *National Museum of the United States Air Force*

This B-26G from the 25th Bombardment Group was painted black on the undersides and up to the level of the bottom of the cockpit canopy, above which the finish was Olive Drab. This squadron was engaged largely in dropping chaff to interfere with enemy radar's ability to detect and track US bomber formations. *Roger Freeman collection*

Martin B-26G-5-MA Marauder, serial number 43-34226, of the 25th Bombardment Group, bore the nickname "Liberty Run." As was the case with the B-26G in the preceding photo, black paint was applied up to the bottom of the cockpit canopy, and a bare-metal area containing the nomenclature and serial number data was preserved just aft of the bombardier's clear nose. *Roger Freeman collection*

The tail and the aft fuselage are the subject of this vintage color photo of B-26G-1-MA, serial number 43-34195. A fair view is available of the Bell M6A tail turret, with a canvas cover instead of the M6 turret's Plexiglas enclosure for the machine guns. *Roger Freeman collection*

"Tail End Charlie," B-26G-25-MA, serial number 44-68254, was the last of the B-26 Marauders. It was delivered to the US Army Air Forces on April 18, 1945, with Martin's chief engineer, William Ebel, piloting the plane for the delivery flight. The "–30–" below the plane's nickname was a newspaperman's symbol for "end of the story." *National Museum of the United States Air Force*

Martin B-26G-11-MA Marauder, serial number 43-34581, is preserved at the National Museum of the United States Air Force. It has been painted to replicate "Shootin' In," B26B-50-MA, serial number 42-95857, which was written off after an accident on April 19, 1945.

The "Shootin' In" replica at the National Museum of the United States Air Force is viewed from the left side. The scoreboard on the forward fuselage, containing nine groups of fifteen yellow bombs, signified that the plane had completed 135 bombing missions. The squadron code was FW and the call letter was K.

"Shootin' In" is displayed indoors at the National Museum of the United States Air Force. The left side of the windscreen was interrupted by a thick, vertical frame and an equally substantial diagonal frame; the diagonal frame separated a fixed-top glass panel and a hinged-bottom glass panel. The bottom panel could be opened to give the pilot a clear view ahead when there were icing problems on the windscreen. *National Museum of the United States Air Force*

A left view of the clear nose on "Shootin' In" includes the heavy-duty frame members that helped support the .50-caliber machine gun and absorbed its recoil during firing. The machine gun was mounted on an E-11 Recoil Adapter, a frame containing recoil buffers and grips.

Martin B-26G-11-MA, serial number 43-34581, the "Shootin' In" re-creation, is observed from the right front in its museum setting. A mannequin representing the bombardier is seated inside the nose. In this photo, the Browning M2 .50-caliber machine gun and E-11 Recoil Adapter have been removed from the socket mount in the nose.

The package guns on the right side of the fuselage of "Shootin' In" are viewed close-up. The pilot fired the guns by using pin-type sights mounted to the front of the windscreen. Ammunition was in boxes inside the fuselage.

The cockpit of "Shootin' In" is pictured. The bombardier gained access to his compartment in the nose through the opening to the front of the copilot's seat, to the right of the instrument panel. The pilot and copilot had control columns (note the mismatched control wheels), but only the pilot had rudder pedals. *National Museum of the United States Air Force*

This is the radio compartment in B-26G-11-MA, serial number 43-34581, facing forward. On the top rack to the front of the radio operator's seat is a pair of SCR-274 Command Radios on a shock mount. The folded-up desk is hiding from view a BC-348-Q Liaison Radio Receiver and a telegraph key. On the left wall are more radio controls, including for the radio compass and the radio operator's intercom. In the bulkhead to the right is the passageway into the cockpit. *National Museum of the United States Air Force*

To the pilot's left side were, left to right, the vacuum-selector control, radio control boxes, an intercom control box, and radio filter control box. Quilted soundproofing material is on the sidewall. *National Museum of the United States Air Force*

To the rear of the radio operator's and navigator's compartment is the forward bomb bay, which is seen here from its right side facing forward. Partially masked by the center bomb racks at the center of the photo is the oblong door into the radio operator's and navigator's compartment. Bomb racks are also on the outer walls of the bomb bay. *National Museum of the United States Air Force*

CHAPTER 4

XB-26H

An unusual experimental aircraft based on a Marauder airframe was the one nicknamed "Middle River Stump Jumper," a TB-26G-25-MA advanced trainer, USAAF serial number 44-68221, redesignated the XB-26H, which was modified with a tandem main landing gear on the fuselage bottom and outrigger gears in the main landing-gear bays. This plane was intended to test the tandem-landing-gear concept that was envisioned for two bombers that were on the drawing boards: the XB-47 and the XB-48. *National Museum of the United States Air Force*

Splines were applied to the fuselage of the XB-26H to strengthen the fuselage. They extended from the sides of the cockpit to about midway between the trailing edges of the wings and the front of the dorsal fin. The outrigger wheels did not always touch ground but were intended to keep the plane stable during takeoffs and landings. *National Museum of the United States Air Force*

The XB-26H had the canopy and side windows of the Bell tail turret, but the turret was not armed, and the rear of it was covered with a pointed fairing. The purpose of mounting tandem main landing gear in the planned XB-47 and XB-48 bombers was because the wings of these jets would be too thin to house the landing gears and their bays. *National Museum of the United States Air Force*

The pointed fairing at the rear of the fuselage of the Martin XB-26H is seen from the right-rear quarter as the plane taxis on a tarmac. Both main landing gears had the wheels and tires on the left side of the strut. Here, the right outrigger wheel is touching the pavement. *National Museum of the United States Air Force*

The XB-26H is being subjected to stringent taxiing tests, with the rear main tire and the right outrigger tire burning rubber. A Martin pilot conducted the test shown in this photo, likely sometime between May and June 1945. *National Museum of the United States Air Force*

The Martin XB-26 is seen in flight during tests. The outrigger landing gears have been retracted into the original main landing-gear bays in the engine nacelles, but the main landing gear remains lowered. The main gear was designed to retract into the fuselage. *National Museum of the United States Air Force*

CHAPTER 5
Production

In 1941, before the United States entered World War II, Chrysler subcontracted to produce B-26 fuselage sections for final assembly by Martin's Omaha plant. This work was undertaken at the Chrysler DeSoto-Warren plant in Detroit and entailed massive retooling by the automaker. Shown here is a forward-fuselage assembly under construction by Chrysler. *National Museum of the United States Air Force*

Chrysler DeSoto-Warren produced forward and center fuselage sections. The forward sections, also called the control sections, included the bombardier's compartment, the cockpit, and the navigator's and radioman's compartment. A row of control sections is under construction in this photo. *Jim Gilmore collection*

The center section was also called the bomb bay section, and one such assembly is seen here from the front end, showing the oblong door from the forward bomb bay to the radioman's and navigator's compartment. At the bottom of the forward bomb bay is part of the massive longeron that gave structural continuity through the open bottoms of the bomb bays. Toward the top of the bomb bay is where the center section of the wing passes through the fuselage. To the rear, the open bottom of the aft bomb bay, later an ammunition compartment, is visible. *National Museum of the United States Air Force*

A workman installs a conduit in the forward left area of the radioman's and navigator's compartment of a B-26 control section under construction at Chrysler DeSoto-Warren. To the man's right is the door to the cockpit. At the top is the round escape hatch for this compartment. To the left is the small window for the radioman. *National Museum of the United States Air Force*

While Chrysler was producing fuselage components for Martin Omaha, Martin's facilities in Middle River, Maryland, were constructing components and assembling them. In addition to existing Plant No. 1 at Middle River, Plant No. 2 was constructed in 1941, not far from Plant No. 1. Shown here are B-26 control sections awaiting final assembly in January 1945. *National Museum of the United States Air Force*

The final-assembly area of Martin Plant No. 2 is seen from an overhead crane on January 24, 1945. To the far right are navigator's and radio operator's sections. To the right of center are control sections, comprising the bombardier's compartment, cockpit, and navigator's and radioman's compartment. To the left of center are center sections, to the far left are tail sections, and in the background are B-26s in the final stages of assembly. *National Museum of the United States Air Force*

In a photo taken closer to the final-assembly line at Martin Plant No. 2, B-26 tail sections (*left*) and center sections (*right*) are in the foreground, with control sections to the far right. In the background, B-26s under final assembly are coming toward the photographer on the right, and then the assembly line makes a 180-degree turn, with the tails of the planes in the left background pointing to the photographer. *National Museum of the United States Air Force*

Another photo of the final-assembly floor at Martin Plant No. 2 shows B-26 control sections in the foreground, and the navigator's and radioman's compartment to the right. Separating these two lines are rows and rows of parts cabinets. In the background to the right is a storage area for components. *National Museum of the United States Air Force*

The Goodyear Aircraft Corporation was another subcontractor for the Martin B-26 medium bomber. At its Plant D in Akron, Ohio, Goodyear built wing sections for the B-26s and then shipped them to Martin's Omaha plant for final assembly on B-26 fuselages. In this photo, wing sections are under construction to the right, and to the left are wing sections with nacelles installed. *National Museum of the United States Air Force*

Wing sections with nacelles in place are viewed from overhead at Goodyear's Plant D in Akron, with one wing toward the left still lacking the nacelle. Visible on each wing are numerous access openings. Goodyear experienced problems in keeping up with production demands for its wings, largely due to engineering concerns related to the transition from the original, shorter wings of the Marauders to the longer wings. *National Museum of the United States Air Force*

Factory employees pose next to the three main fuselage sections of a B-26 before they are fastened together. A sign on the center section proclaims that this is a special fuselage, but only the words "The Last Sections of the" and "B-26" are visible. Two factory numbers are stenciled on the nose: F456 and G456. *National Museum of the United States Air Force*

Fuselage assemblies are installed on trolleys for easy movement from assembly station to station at Martin's Middle River Plant No. 2 in late January 1945. The canopies are present on the tail turrets on the line of fuselages in the foreground. In the background are B-26s with the wings and empennages installed. *National Museum of the United States Air Force*

Around the beginning of 1942, a mechanic checks clearances as a Pratt & Whitney R-2800 Double Wasp engine assembly is being maneuvered into place on a B-26 nacelle. He has his right hand on the filler tube of the oil tank and his left hand on the tubular engine support, which will be attached to an engine mount that is fastened to the wing structure. *National Museum of the United States Air Force*

Engines are installed on these B-26s on the final-assembly line. They already have received their national insignias on the wings: the late type with the side bars. Bomb bay doors, empennages, top turrets with machine guns, and package gun fairings are installed. Propellers and cowls are yet to be mounted. *National Museum of the United States Air Force*

In a general view of the final-assembly area, note on the nearest plane the protective covering, possibly frisket paper, on the Plexiglas canopy and rear dome of the tail turret. The doped fabric of the rudder and elevators contrasts with the highly polished Alclad aluminum-alloy skin of the dorsal fin and horizontal stabilizers. *National Museum of the United States Air Force*

Protective padded mats were laid on the wings of the B-26s on the assembly line to protect the Alclad skin from scuffs, spills, and damage. Liquid containers are on the right wing of the B-26 in the foreground. Note the open, round hatch at the top of the navigator's and radioman's compartment. *National Museum of the United States Air Force*

At a turnaround in the final-assembly line, the B-26 to the right has a fixed .50-caliber machine gun mounted in the bombardier's compartment. Later, the Plexiglas nose will fit over this gun's barrel. The pilot's ground-strafing armor has been installed on this plane, and the package machine guns and their fairings are present. *National Museum of the United States Air Force*

The B-26 in the foreground has the early-type hatch for the life raft, which also served as an extra escape hatch, and this was offset slightly to the left of the longitudinal centerline of the top deck of the fuselage. Between the top turret and the dorsal fin are three formation lights; two more formation lights are visible on the left horizontal stabilizer, and the right horizontal stabilizer also had two such lights. These lights had blue lenses of teardrop shape. *National Museum of the United States Air Force*

A Martin employee updates her checklist under the right wing of a B-26 nearing completion on December 31, 1943. The nose-gear wheel and tire are wrapped in protective material and shackled to a trolley mounted on tracks on the factory floor. The main-gear wheels are covered with fabric shrouds. Propellers are yet to be mounted on these planes. Note the opening in the clear nose of the closest plane for a fixed .50-caliber machine gun, along with the ball mount for a flexible .50-caliber machine gun at the tip of the nose. *National Museum of the United States Air Force*

Workers at Martin's Middle River, Maryland, plant are assembling early B-26s, decorated with prewar red, white, and blue stripes on the tails. Although the B-26 started its career with the reputation of a dangerous and unpredictable plane to fly—a widowmaker—the plane eventually proved its worth and was a major contributor to the United States' tactical airpower in World War II.
Air Force Historical Research Agency